May 1

We're off to settle new land out west! Our wagon is packed with big and small things.

We brought all that could fit inside! We brought food, tools, bedding, even books to study. We're going to a new land, a new home.

May 3

Father was voted wagon master for the wagon train. All the men thought he could lead us well.

Father is proud, but says the job is hard. He must get us over the mountains by fall. Snowfall on the mountains can bring great danger.

June 1

We have been on the trail for one month. It's hot and dusty and I'm getting very tired. Each day all we do is walk beside the wagon. Strange, I always thought wagons were for riding in! Father says even large mules can't carry everything!

June 20

Walk, walk, walk, always walking, all the time!

I told Father I wanted to stop and play. I was so cross, I almost fought with him. Father had a small talk with me then. "Daughter, listen, it's important we hurry," he told me.

Father said we must get to the mountains soon. Or early fall snow could block our trail!

"Daughter, this trip is hard for all," he said. "But one day we will laugh and be glad. We will have our own land, our own small farm!"

July 3

Yesterday noon I heard Father call out, "Mountains!" Father told us he thought we were making good time.

Last night there was a happy party! Father played his fiddle as we sat round the fire. There were songs, jokes, and many a laugh also!

July 20

I never thought mountain trails would be so bumpy! Today a wheel broke going over a rock. A new one cannot be bought in the mountains. So Father sawed up a table to make one. Mother fought back tears at seeing her table.

September 1

Today, early, I laugh and call out, "Almost there!" We have crossed the country and ridden over the mountains! No one can talk of anything else now! We're almost there ... almost home!